HOT ON THE TRAIL

The only Alaskan wildlife guide
you'll ever need

An almost serious approach
to identifying that which has travelled
the trail before you

L.L. Simons

Copyright © 1993 by Cold Country Publishing.
All rights reserved. No part of this book may be reproduced in any form, except for brief reviews, without the written permission of the publisher.

Cold Country Publishing
P.O. Box 22498
Juneau, Alaska 99802-2498

First Printing 1993

This book was produced on a NORCOM 486 PC clone using Pagemaker 4.0. Camera ready copy was produced on a HP LaserJet 4 using Bodacious (a truetype font).

Cover, layout, and typography
      by David Riccio
Inside illustrations
      by L.L. Simons
Photography
      by L.L. Simons

Library of Congress Catalog Card Number: 92-74442

ISBN 0-9634877-0-1

Manufactured in the United States of America

ACKNOWLEDGEMENTS

Wayne Ray and his staff at the Anchorage Zoo deserve special recognition for their assistance. I certainly cannot thank them enough for their pleasant cooperation. They didn't even laugh when I told them what I wanted to do.

I also want to thank Glen. He laughed a lot when I told him what I wanted to do.

CONTENTS

Introduction	vii
How to Find Wild Animals in Alaska	1
Moose	5
Caribou	11
Musk Oxen	15
Sitka Black Tailed Deer	19
Bear	23
Porcupine	27
Mink	31
Snowshoe Hare	35
Fox	41
Wolf	47
Dall Sheep	51
Ptarmigan	55
Call of the Wild	59
Joel's Pancakes	63
An Alaskan Fable	65

INTRODUCTION

When I was young and looking for adventure, I wanted to come to Alaska just for a visit. In the spring of 1978, I read a classified ad in <u>Mother Earth News</u> saying that two women were willing to drive anyone's car to Alaska. The phone number was in my Michigan area code, so I called and told the lady who answered that I did not have a car, but would like to go with them.

Becky and I met and got along famously. She consulted with her travelling partner in Chicago, and we unanimously decided that it would work out just dandy for the three of us to travel to Alaska together. We never did find a car to transport, so we planned to make our trip by plane, ferry, foot and thumb.

Two days before we were to leave, Becky called and said she had some problems that could not be resolved in time to make the journey. For her, the trip was off. I called her friend, Sherry, in Chicago and told her that I was going to Alaska, even if I went alone, and would she like to join me? I left for Chicago the next morning, spent the night at Sherry's apartment, and with her, boarded a plane for Seattle the next day, headed for Alaska.

Sherry and I had a fine trip. We covered and slept on a lot of Alaskan ground in the month we had. It was on this, my first trip to the state, that I became keenly interested in what I was finding on the trail. It was like a "who done it?" mystery. I decided that a scat guide would be just the thing for a cheechako like me (that's Alaskan for "Arctic idiot").

So here it is, the book we've all been waiting for, fifteen years in the making. I'm hoping to expand my photo collection in future versions, but for now, I think this will get you down the trail and back in most parts of Alaska.

Happy travels.

HOW TO FIND WILD ANIMALS IN ALASKA

First, if you are not here already, you must actually visit Alaska. Get as far away from a city and as close to a national park as you can. In general, these simple rules of the road will prevail:

1. Go where the animals are. Although moose live within the Anchorage metropolitan area and black bear cruise the streets of Juneau at night, you normally will need to go outdoors to see wildlife. The more time you spend out in the wilderness, the more animals you will see.

2. You have to be QUIET. Rid yourself of talkative companions, walkmans, and anything that jingle jangle jingles. Pick up your feet when you walk. But BEWARE - Going quietly through the woods in search of animals is also called stalking. Some animals, such as bear, do not take kindly to stalking. Try to bring along a friend who is slower and weaker than you are in case you have a bear encounter.

3. Visit national parks and wildlife refuges. This is where the action is, and where there are at least some land and water trails to take you through the wilderness. Many areas of Alaska can be hard to traverse due to lack of adequate

trails. Except in popular areas, the word "trail" can be a misleading one. All it really means in Alaska is that someone or something has gone this way before, but it doesn't necessarily mean that they survived the trip. If you do much remote hiking here, you will remember these words.

4. Take the time to stop, look, and listen. Do not forget that you live in a three-dimensional world. Look around and above you. Try to spend time sitting quietly, listening, and observing. You will be surprised at what you see. It's a jungle out there.

5. Non-motorized vehicles such as canoes, kayaks, rowboats and rafts are wonderful for viewing wildlife on ocean, lake, or river shores.

6. Although a seeming contradiction to all the suggestions above, a flight-seeing trip can be the ultimate animal viewing adventure. You won't be outside, and you cannot possibly make more noise than an airplane. But it increases your field of vision, and animals are not especially frightened of planes unless they fly too close. Book a flight that is specifically meant for animal spotting, because even in Alaska some areas are relatively devoid of viewable wildlife.

NOTES

Moose

6

Moose Nuggets

Moose nuggets are so distinctive that the folks in Talkeetna, Alaska hold an annual Moose Dropping Festival. Seems that an animal rights group became a little concerned, however, when they read about it. They wanted to know how high up the moose were held before being dropped.

Moose graze in woodland ponds, swamps, and lakes.

Range: Most of Alaska, except some offshore islands and the far Southwest.

MORE MOOSE

Did you ever wonder what the Creation Committee was thinking about when they designed a moose? Here is an animal that stands over 7 feet tall at the shoulder. That puts myself and other humans of shorter stature about eye level with moose knees. Not a pretty sight. On top of these four wobbly knees is 1,000 pounds of lumpy, hairy body with a protuberant snout that is often dripping with swamp debris. The icing on the cake of this lovely specimen is a hat rack made for Goliath, measuring up to five feet across, that you do not want to get into an argument with. Certainly a visualization of the word "beast".

Moose have found their way into the hearts of most Alaskans via the hearty and not-too-fussy Alaskan stomach. It seems to be a rule of the wild that the worse something looks, the better it tastes, and the mighty moose is no exception. For people in remote areas of Alaska, it is standard and inexhaustible winter fare. They steak it, chili it, jerky it, and fry it. They corn it, soup it, and burger it. And still the freezer is full of moose - This is a big animal! So they roast it and barbeque it, hash it and grill it. Moose tacos, moose pizza, moose spaghetti, moose sausage. There is no end to

the culinary delights that have been devised for trying to face, once again, another moose meal at the end of a long, isolated winter. Saving the best for last, there is also the lovely dish of moose-nose soup, which is a delicacy in parts of Alaska that I hope never to visit.

NOTES

Caribou

Cari-Bou-Boos

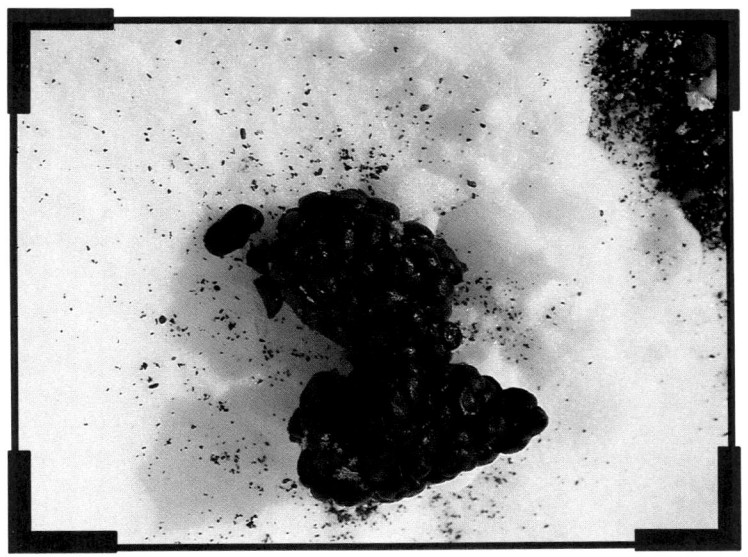

A wild version of Rudolph, Prancer, and the ho-ho-ho gang. These transients of the tundra move in vast herds, literally a cast of thousands. They set a liberated example, as both male and female caribou have antlers.

Range: Most of Alaska except some offshore islands and Southeast Alaska.

NOTES

Musk Oxen

Musk Ox Mounds

Dumb as a rock, they stand in a circle with heads out when threatened. Kinda like old John Wayne in the wagon train movies. Both of them got shot to hell. Musk ox were reintroduced to Alaska in the 1930's.

Range: Wild herds exist only on Nunivak and Nelson Islands, and on the fringes of the far western arctic slope.

NOTES

Sitka Black Tailed Deer

 20

Deer Dumplings

According to biologists, the average deer makes a deposit thirteen times in twenty-four hours. Supposedly, you can count the number of fresh dumplings in an area, divide by 13, and know how many deer are watching you and laughing. Remember this little activity if the kids get bored camping.

Deer are woodland creatures, commonly sighted along ocean shorelines, especially at low tide.

Range: Southeast Alaska, Kodiak Island, Afognak Island, Prince William Sound, and Yakutat area.

NOTES

Bears

Brown

Black

Polar

 24

Bear Blob

Bear in Alaska come in three flavors: black, brown, and polar, the order of their size and temper. Many theories exist on how to react if attacked. My favorite piece of advice is to place your feet in a wide stance, bend at the waist, put your head between your legs, and kiss your butt good-bye.

Black and brown bear frequent salmon spawning streams and berry patches. Polar bear are found on the far northern ice pack and are considered a marine mammal.

Range: Wherever you are in Alaska, there is probably a bear nearby.

NOTES

Porcupine

28

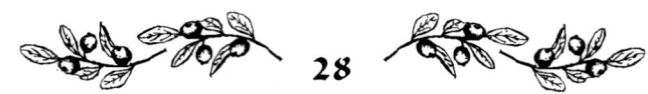

Porky Pucky

An odd creature, make no mistake. Just think of the things we take for granted that must be darn difficult for porcupines. How, you may wonder, do they mate? Very carefully, no doubt. It makes me shudder just to think about them giving birth. Youch!

Range: Wooded areas throughout Alaska, except for most coastal islands.

NOTES

Mink

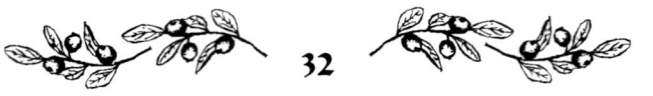

Mink Manure

These feisty members of the weasel family are entertaining to watch, especially when it is someone else's food cache they are pilfering. Mink are not socially acceptable in some communities, since they are strict carnivores and insist on wearing only real fur.

Mink like to be near water, either salt or fresh.

Range: Throughout Alaska except the Arctic Slope and some offshore islands.

NOTES

Snowshoe Hare

36

Bunny Buttons

Like the rest of the rabbit family, the snowshoe hare spends half of its life hiding from predators. And we all know what rabbits do with the other half of their life. I guess you'd call it a "good news, bad news" existence.

The snowshoe hare likes wooded or brushy areas, the better to hide in, I presume.

Range: All but far Western Alaska, the North Slope, and coastal islands.

THE HARE FACTS

The question is, why is it a snowshoe hare and not a snowshoe rabbit? You have undoubtedly heard it both ways, but now the truth can be told. It is all in how they are born. Hares, which include jackrabbits, are born with hair (didn't you just know there was a connection?), with eyes open and good mobility within minutes of birth. Rabbits, like humans, are born naked and helpless. There is no such thing as a snowshoe rabbit, except that some people are stuck on the term and apply it to our very own snowshoe hare. Got it?

No, you don't. There is also an Alaskan hare, which is not a snowshoe hare. Also referred to as an arctic hare, it turns white in the winter just like its bigfoot cousin the snowshoe. The Alaskan hare inhabits the coastal areas of far western and northern Alaska, while the snowshoe hare occupies the interior and the southern and eastern coasts.

NOTES

Fox

Fox Fudgies

Like the bears, they come in different colors: red, gray, and arctic white. The classic fox hunt never caught on in Alaska. It is hard to look sophisticated on a snow machine in a parka. And I suspect maybe your lips would stick to the bugle at $40°$ below zero.

Range: Throughout the state except for Southeast Alaska, coastal islands, and the Aleutians.

FOX ON THE RUN

When a fox wanders into a remote Akaskan village, it is a cause for concern among the townfolk, for such a brave showing is often an indication of rabies. In days of not so long ago, most villages had no telephones, and it was part of the daily acitivity to keep the CB radio on so that you could talk to your neighbors and hear what everyone else was talking about too. Life held few secrets.

On a cold spring day in a Bristol Bay village in the early 1970's, Mr. Fox came to town. He was announced by an ominous voice on the radio declaring to all who were listening, "Rabie fox headed for Mary's house".

Because it was spring, all the men were at the local cannery preparing for the start of the fishing season, and the children were in school. Mr. Fox was at the mercy of the village ladies. The CB rattled with activity as the women went into action to protect their homes from the marauding red beast.

"I cannot find gun."
"Fox is heading for Annie's house."
"The gun is outside, I'm not going to go get it."
"He's leaving Annie's, headed for Ella's house."

"I got gun, but no shells."
"Fox is heading for Nina's house."
"I got shells, what kind gun you got?"
"Fox is heading for my house."
"You shoot him."
"Fox is on Sophie's porch."
"He'll be safe there, she can hit nothing."
"Do you think he can get in my house?"
"I cannot see fox no more."
"There he goes."

  Outside, the streets were quiet except for the loping fox, picking his way past the low-roofed, smoky houses. No one ever knew whether the fox was rabid or just curious. By the time the right gun got into the right hands with the right shells, Mr. Fox was already into his next adventure of the day, unscathed. In his wake he left the village, with its twenty or so houses scattered along the one-mile path, to look like a little tornado had come through only on the inside. I bet they were ready next time.

NOTES

Wolf

Wolf Winky

Alaska's gray wolves are the largest of the dog family. Wolves even look like big dogs, but they are more often heard than seen. Our summer visitors miss the experience of standing outside listening to the howl of the wolf on a frozen, still night with the northern lights sizzling overhead. If you're just passing through, come back and see us in the winter.

Range: Just about everywhere in Alaska.

NOTES

Dall Sheep

Dall Sheep Doo Doo

Only hearty hikers and white-knuckle fliers will view these beasties in their habitat, hugging a mountain at 5,000 feet or more. Head butting seems to be a favorite passtime.

Range: Brooks, Alaska, Wrangell, Talkeetna, and Chugach mountain ranges, as well as the Tanana-Yukon uplands.

NOTES

Ptarmigan

Ptarmigan Pturds

> OK, there's no such thing,
> but I couldn't resist the phrase.

There's a town in Alaska called Chicken. Rumor has it that the town got its name because nobody there could spell "ptarmigan".

You will find Willow, Rock, and White-tailed ptarmigan.

Look for them in high, treeless areas.

Range: Throughout Alaska.

NOTES

CALL OF THE WILD

Alaskans love wildlife. They like to look at it, paint it, eat it, and tell stories about it. I guess it makes sense, then, that when choosing a business name, animals are at the top of the list.

Southeast Alaskans seem especially fond of bear. Down there you can stay at Glacier Bear Lodge, rent a movie from Bear Video, buy snacks at a grocery store called Super Bear, and fix your dents at Bear Body Works. Fill 'er up at the Black Bear Quickstop, and get Brown Bear Cleaning Services to make your home or office look like new. Keep those pipes clean and the home fires burning with Polar Bear Plumbing & Heating.

Moving on, I found the game name to be popular in fine dining establishments throughout Alaska. For example, you can have yourself a good snack and a great time at the Fox Roadhouse. Order prime rib, crab, and other delicacies at the Bear and Seal Restaurant in Fairbanks. (This name makes me a little nervous. Would you eat at a restaurant in Mexico named the Cat and Dog?) The Moose Restaurant in Petersburg is not at all related to Bullwinkle's Pizza Parlour in Juneau. And things have gone to the birds at the Raven

Roost. Don't miss Porcupine Pete's in Haines, but if you want something a little lighter, stop by the Lynx A&W in Anchorage for a hot dog and a root beer.

Like Alaska so much you want to stay? Run out and buy your own piece of the Last Frontier from Walrus Properties in Anchorage. Then get in touch with Musk Ox Super Insulated Homes and have them build your dream. Buy your lumber from Black Bear Cedar Products in Thorne Bay, and have Little Fox Construction in Sitka arrange for your roof. A Glacier Bear Fence will keep the moose out of the potato patch, and you can finance it all at the Caribou Loan Company in Fairbanks. And let's not forget to insure with Eagle Pacific.

What we all want to know is, what do they make at the Moose Nugget Factory in Talkeetna? And just HOW do they make it? This may be a little technical, but I especially want to know how they calculate their work-in-process inventory. Do people really list these places on their resumes?

Sasquatch is alive and well and living in Alaska. You can go to Big Foot Auto Service in Haines, but when your pipes are frozen, who ya' gonna call? Bigfoot Pumping and Thawing on Loose Moose Loop in Fairbanks. Try that ten times fast. I think they should relocate the factory in Talkeetna to this address, or at least make a serious effort to find the road's namesake and boost their profits.

Try to decide who views who at the Eagle View Car Wash in Palmer. Then take that clean car down the road to Double Eagle Real Estate and Investments. That should be about all the eagles a person can handle in Palmer, Alaska in one day. BUT WAIT...before you go, be sure to drop off your baby's diapers at Bear Buns Diaper Service.

What's in a name? In Alaska, apparently everything furry or feathered.

JOEL'S PANCAKES

Joel came to Alaska to work with us for the summer at a remote fishing lodge on Lake Clark, in Southcentral Alaska. One of his first assignments was to set up an outpost camp on a fly-in river. Joel was dropped off with the gear he would need to set up a tent frame (a cross between a tent and a cabin) and make the place comfy. We left him there alone, and came back to pick him up a few days later.

Joel was enthusiastic about his solo adventure in the Alaska wilderness, and he told us all that he saw and did out there by his lonesome. He talked of the fresh fish he ate and the terrific blueberry pancakes he made himself for breakfast.

But we said, "Gee, Joel, where did you get blueberries this time of year? It's only May. They don't come out until August." Joel said he didn't know about all that, but he sure enough found a big pile of blueberries right on the ground. Come to think of it, though, he had noticed that they were the only blueberries he'd seen.

Slowly, those of us listening started to smile as we realized the source of Joel's blueberries. The smiles turned into laughs, then guffaws as knowing looks passed back and forth.

Now, it's one of life's mysteries as to why a bear eats berries. They come out of the bear looking no different than when they went in. Over the frozen Alaska winter, as the elements of nature claim the digested remains left by brer bear in the fall, they leave intact those stubbornly unaltered berries.

We explained to Joel how the blueberries on the vine in August can end up in a neat pile on the ground in May. He didn't think it was nearly as funny as we did. We noticed he kind of lost his taste for berries after that.

AN ALASKAN FABLE

Once upon a time there was a nonconforming arctic tern who decided not to fly south for the winter. However, soon the weather turned so cold that he reluctantly started to fly south. In a short time ice began to form on his wings, and he fell to earth on the tundra, almost frozen. A caribou passed by and crapped on the little arctic tern. The arctic tern thought it was the end. But the manure warmed him and defrosted his wings. Warm and happy, able to breath, he started to sing. Just then, a lynx came by and, hearing the chirping, investigated the sounds. The lynx cleared away the manure, found the chirping bird, and promptly ate him.

The moral of the story:

Everyone who craps on you is not necessarily your enemy.

Everyone who gets you out of the crap is not necessarily your friend.

And if you're warm and happy in a pile of crap, keep your mouth shut.

NOTES

ABOUT THE AUTHOR

L. L. Simons spent eleven years assisting in the operation of remote sport fishing lodges both at Lake Clark and Prince of Wales Island. She currently resides in Juneau, Alaska. This is her first book.

NOTES

More about Cold Country Publishing:

This is the first release by Cold Country Publishing. We all have to start somewhere. <u>Hot on the Trail</u> is available at bookstores and at Alaskana gift and curio shops. You can also order direct from the publisher. The price of the book is $6.95. Please add $1.25 for shipping and handling for the first book, and 50 cents for each additional book. Juneau residents include local sales tax.

Cold Country Publishing

P.O. Box 22498
Juneau, Alaska 99802-2498